Boroleas The Alchemical Wizard With Boron Magic

Element 5 – Magical Elementals Chemistry Fantasy

By Sybrina Durant

Boroleas The Alchemical Wizard With Boron Magic

Element 5 – Magical Elementals Chemistry Fantasy

Paperback ISBN 978-1-942740-74-2

BISAC Codes:
JUV036000 - JUVENILE FICTION / Science & Technology
JUV037000 - JUVENILE FICTION / Science Fiction, Fantasy, Magic
JUV002270 JUVENILE FICTION / Fantasy / Dragons, Unicorns & Mythical Creatures

League City, Texas, United States of America

Contact Sybrina@sybrina.com.

No Metal
Boroleas
Boroleas The Wizard With
The Boron Staff
B
No Magic

Chapter 1: The Glowing Sigils of Crystalmere

The market district of Crystalmere had not changed much in three hundred years. Brass-fitted gas lamps lined cobblestone streets where vendors sold everything from alchemical reagents sealed in crystalline vials to neural-linked communication scrolls that buzzed with encrypted whispers. Steam-powered trams rattled alongside horse-drawn carts. Holographic sigil-boards advertised reagent merchants beside hand-painted tavern signs. MarBryn was like that — ancient and modern, tangled together like copper wire through old stone walls.

Boroleas moved through the crowd with the quiet authority of someone who owned whatever room she entered. Her blue-black hair caught the amber glow of the gas lamps, and her deep magenta hood billowed at her shoulders as though some invisible current perpetually followed her. Beneath the hood, her violet-blue eyes moved steadily across the market — scanning, cataloguing, never missing a detail. The golden alchemical sigils stitched along the hem of her black robe caught the light with each stride, and the circular blue gemstone brooch at her chest pulsed faintly, almost like a second heartbeat. The pale blue-grey gloves on her hands flexed slightly as her fingers moved, a habit born from years of keeping her magic ready.

She was not browsing.

She was following a trail of energy. Specifically, a trail of destabilized boron molecules — and only someone with her particular sensitivity to Element 5 would have noticed it at all.

"You have that look again," said a voice from somewhere near her left knee.

Boroleas glanced down. Prism, her companion, trotted alongside her on four crystalline legs that clicked rhythmically against the cobblestones. He was a Luminal Fox — a rare magical creature native to the frost caves of the Aetherveil Mountains — and his semi-transparent body refracted light in small rainbow fractals wherever he stepped. His eyes were sharp silver, always slightly amused. He stood roughly the size of a large house cat and had the attitude of someone who had lived at least four hundred years. He probably had.

"What look?" Boroleas said without slowing.

"The look where your left eyebrow goes precisely three millimeters higher than your right and you start curling your gloved fingers. Imminent trouble. I've catalogued seventeen separate instances."

"Eighteen now," she said. "Something is wrong with the Boron lattice beneath the market square. Someone has been tampering with the foundational matrix."

Prism's silver eyes narrowed. "The alchemy grid? That grid has been locked since the Second Concordat."

"Exactly."

The alchemy grid was not common knowledge. Most residents of Crystalmere knew only that their city ran clean and warm, that the water filtration never failed, that the structural ceramic of their buildings never cracked regardless of seismic tremors. What they did not know was that beneath the city, embedded into the foundation stones by the ancient Concordat engineers, was a lattice network of boron-ceramic conduits — a masterwork of alchemical engineering that stabilized the city's molecular infrastructure. Boroleas had helped reinforce it herself, nearly sixty years ago. She did not look eighty. She did not feel it either.

She stopped at a market stall selling bulk powdered reagents. An elderly man with ink-stained fingers peered up at her over enormous spectacles.

"Alchemist Boroleas," he said, straightening. "Haven't seen you since the Glasswright Festival."

"Alderman Petch. Have you noticed anything unusual in the square these past three days? Vibrations, perhaps. Unusual heat from the pavement?"

Petch scratched his jaw. "Now that you mention it, my reagent scales have been off. Calibrated them twice. And my neighbor — the ceramics woman two stalls down — she said her display cases cracked in the night. Clean right through. No impact. Just cracked."

Boroleas and Prism exchanged a glance.

"Ceramic fractures with no impact point," Prism said quietly, "suggests a resonant frequency disruption from below. Someone is running an unauthorized extraction frequency through the grid."

"Draining it," Boroleas said flatly.

She pressed one pale blue-grey glove to the cobblestones, closed her violet-blue eyes, and reached inward — past stone, past mortar, past centuries of compressed mineral — until she found the familiar crystalline hum of the boron lattice. She let her awareness travel along it the way a musician traces a string. And then she felt it. A wound in the matrix. A deliberate cut, surgical and cold, like something had been feeding on the grid's energy from a specific point.

She stood, her expression unreadable.

"Someone is stealing the city's structural foundation," she said. "If the lattice fails, every boron-ceramic building in Crystalmere comes down."

Prism sat back on his crystalline haunches. "Well. I was hoping we might have a quiet week."

"When," said Boroleas, already moving, "have we ever had a quiet week?"

Chapter 2: The Merchant Who Knew Too Much

The address came from Alderman Petch — a name he gave reluctantly, in a lowered voice, after Boroleas placed two gold alchemical marks on his counter and held his gaze long enough that he understood she was not asking. The name was Drevon Salk, a materials merchant operating from a converted warehouse in the Ironholm district, the oldest quarter of Crystalmere, where the buildings were black with three centuries of foundry smoke and the streets were too narrow for the trams.

Drevon Salk, it turned out, was not what he appeared to be.

He presented as a compact, well-dressed man in his middle years, with the practiced sociability of someone who sold things for a living — warm handshake, easy smile, a workshop full of legitimate boron-alloy components and certified chemical documentation posted on a smart-board beside his desk. The workshop smelled of hot metal and industrial flux. A young female apprentice worked a press in the back corner, studiously not looking up.

"Alchemist Boroleas," Salk said, spreading his hands welcomingly. "What an honor. How can I assist?"

"You can tell me where you are routing the extracted lattice energy," Boroleas said pleasantly.

The smile stayed on his face, but something behind his eyes changed.

"I'm afraid I don't follow."

Prism hopped onto a workbench and examined a rack of boron nitride rods with what appeared to be casual curiosity. "These rods have been recently exposed to a resonant extraction field," he said conversationally. "You can see the micro-fracturing at the crystalline boundary. Also, your wall clock has stopped, which suggests a significant electromagnetic event in this room within the last twenty-four hours. Also, your coffee cup has a stress crack consistent with thermal shock from an unshielded energy discharge." He looked at Salk with silver eyes. "Also, you're sweating."

Salk dropped the smile. "I want you to leave."

"I don't want to leave," Boroleas said calmly. "I want to understand who you work for and what they plan to do with energy sufficient to destabilize a city foundation."

The apprentice in the back had stopped working. Boroleas noted that without turning her head. She also noted that three figures had appeared in the doorway behind her — large, wearing the non-reflective composite armor favored by private enforcement units. Each carried a short-pulse discharge weapon. Old military surplus, she assessed, but functional.

"Escort them out," Salk said.

The three moved forward, and Boroleas raised her left hand.

The luminous blue-white orb that formed between her gloved fingers was quiet at first — a soft, swirling glow barely larger than an apple. It expanded slowly, deliberately, until the workshop was bathed in cold blue light and the boron-alloy rods on Salk's shelves were vibrating audibly.

"I would really rather talk," she said.

The three enforcers looked at one another. One of them raised his weapon anyway. Boroleas sent a compressed boron bolt — not large, not intended to harm, but precise — directly into the barrel of the weapon, fusing the discharge mechanism solid. The man dropped it with a shout, shaking his hand.

"Next one goes through the wall," she said evenly. "And the one after that dismantles the structural joints of this building. Which, unlike those on the south side of the square, are not reinforced with boron ceramic and will come apart quite quickly."

There was a long silence.

"Sit down," said the young woman in the back. Everyone turned. She had removed her work apron, and beneath it she wore the close-fitted tactical jacket of a Crystalmere Intelligence operative. She produced a badge — genuine, Boroleas noted, not replica. "Operative Minn Calder. I've been watching Salk's operation for six weeks. I was hoping you wouldn't blow my cover quite so dramatically."

Prism tilted his crystalline head. "You might have introduced yourself earlier."

"I was gathering evidence."

"We were about to be escorted into the street by armed men."

"I had it managed," Minn said firmly, though her expression suggested she was perhaps forty percent confident in that statement. She was young — mid-twenties, Boroleas judged, with close-cropped brown hair and the alert, contained energy of someone trained to read exits before they read faces. She was clearly skilled. She was clearly also, at this precise moment, improvising.

Salk made a move toward a hidden panel behind his desk. Boroleas reached out with her awareness — felt the boron composite of the panel latch — and locked it solid with a thought. Salk's hand hit the panel and it didn't move.

"Sit down, Mister Salk," Boroleas said again.

This time, he did.

Minn secured the enforcers with restraint ties from her jacket, moving efficiently and without drama. Then she pulled a data scroll from her pocket and handed it to Boroleas. The alchemist unrolled it. Her violet-blue eyes moved quickly across the projected information.

"The energy is being routed through underground conduit to a receiving station outside the city walls," Minn said. "We don't know where it goes from there. Salk won't talk. He's terrified of whoever contracted him."

"I can tell you one thing," Salk said quietly, all performance now completely gone. He looked tired and genuinely frightened. "The man who contracted me — he doesn't want the energy to power anything." He paused. "He wants to use it to unmake something."

Boroleas looked at him steadily.

"Which something?" she asked.

Salk swallowed. "The Concordat Vaults. The old ones. Underneath Ashenmoor."

The name landed in the room like a stone dropped into still water. Even Prism went very still.

Minn looked between them. "What are the Concordat Vaults?"

"Old enough," said Boroleas carefully, "that most people believe they are legend." She rolled the data scroll closed. "They are not."

Chapter 3: The Road to Ashenmoor

They left Crystalmere before dawn.

Minn Calder had requested exactly four hours to file a partial report with her intelligence office — omitting, at Boroleas's suggestion, the part about the Concordat Vaults, since the fewer people who knew where they were going and why, the better. She filed the report, packed a field bag with alarming efficiency, and presented herself at the city's eastern gate at precisely the agreed hour, wearing the same tactical jacket and carrying a long-barreled chemical-combustion rifle across her back.

"You travel light," Prism observed.

"I travel smart," Minn said. "There's a difference." She looked at the Luminal Fox with the slightly uncertain expression most people wore when confronted with a crystalline four-legged creature who spoke in full sentences. "Do you — do you need anything? Food? Water?"

"I subsist on ambient luminous energy," Prism said. "But thank you for asking. Most people don't."

The road east from Crystalmere wound through the Thornfeld Plains — a broad expanse of amber grasses interrupted by the occasional outcropping of dark basalt rock and the skeletal silhouette of an old relay tower, its signal dishes long since dead and rusted. The plains stretched flat and wide beneath a sky that was lavender at the horizon and deepening to indigo overhead. MarBryn's moon, was still visible in the early morning sky.

Boroleas walked at the front of the small group, her magenta cape catching the pre-dawn wind, her eyes moving across the landscape in that steady, cataloguing way. She was telling Minn about the Concordat Vaults — or rather, telling her enough.

"The Second Concordat," she began, "was an agreement between the six founding alchemist houses and the early engineers of what became Crystalmere. They were the ones who built the boron lattice grid beneath the city. But before that, they built something older. The Vaults beneath Ashenmoor were a repository — a sealed archive of alchemical knowledge, raw elemental materials, and what the Concordat called Primary Bindings."

"What's a Primary Binding?" Minn asked.

"Think of it as the alchemical equivalent of a master code," Prism said. "A sequence of elemental configurations that, if properly understood and applied, allows its holder to alter the fundamental molecular composition of large areas of the physical world."

Minn absorbed this. "So if someone unlocked those Vaults—"

"They could, with sufficient knowledge and energy," Boroleas said, "rewrite the material structure of anything within range. Stone to gas. Metal to sand. A living city to empty field." She paused. "Or a living person to something considerably less organized."

"That is," Minn said carefully, "an absolutely terrifying thing to have locked under a ruin."

"Yes," Boroleas agreed. "That's rather why it was locked."

By midday they had crossed the plains and reached the edge of the Veldrak Wetlands — a vast, fog-prone region of shallow lakes, reed beds, and floating peat islands, threaded through with narrow elevated pathways that the locals called the Causeways. The fog rolled in low and thick, smelling of minerals and damp moss, and the sound of the world narrowed down to footsteps on stone and the distant, haunting call of something large moving through the reeds.

It was on the Causeways that they found Tessek.

He was sitting on a collapsed section of pathway with his boots off, examining a significant cut on his left forearm with the resigned expression of someone who was accustomed to arriving places slightly damaged. He was a large man — broad-shouldered, somewhere in his late thirties, with dark reddish skin, a close-cropped beard, and the kind of calloused hands that spoke of a life spent working with physical materials. Around his neck hung a collection of small tools on a leather cord, and his belt held a remarkable array of compact devices. He was a Pathwright — one of MarBryn's ancient engineering-navigator class, people trained in the old and new sciences of physical structure, route-finding, and materials engineering.

"You're bleeding on the Causeway," Minn noted as they approached.

"Astute observation," Tessek said without looking up. "I was ambushed by three individuals in composite armor about a mile back. I dealt with two of them. The third was faster than anticipated." He looked up. His eyes registered Boroleas with immediate recognition. "Alchemist Boroleas. You're heading to Ashenmoor."

"How do you know that?" Boroleas asked.

"Because the men who ambushed me asked if I'd seen a woman in a magenta cloak heading east. I told them no, which was true at the time." He began wrapping his forearm with practiced efficiency using a strip of treated fiber from his pack. "I'm Tessek. Pathwright, third rank. I was hired six weeks ago by a private commission to survey and map the underground passages beneath Ashenmoor. I thought it was a structural assessment contract. I was wrong."

Boroleas crouched to his eye level. "Who hired you?"

Tessek looked at her steadily. "A man named Vorrath Keln."

The name was one Boroleas had not heard spoken aloud in over twenty years. But she knew it. She knew it the way you know the specific timbre of a sound that once frightened you badly.

Her violet-blue eyes were very still.

"Then we need to move faster," she said.

Chapter 4: The Ruins of Ashenmoor

Ashenmoor was not a place that invited visitors. The ruins rose from the edge of the wetlands like broken teeth — dark stone structures, half-submerged in peat, their original architecture barely discernible beneath centuries of root intrusion and chemical weathering. The sky above the ruins was perpetually overcast, and the air carried a faint metallic sharpness that set the teeth on edge. Boroleas recognized it immediately: trace boron particulate, drifting up from the disturbed Vault passages below.

"Someone has already been excavating," Tessek said, reading the ground with the practiced eye of a Pathwright. He moved carefully between the stones, touching surfaces, checking angles. "These excavation cuts are fresh. Industrial-grade sonic boring equipment, based on the residue pattern."

"How far down are the Vaults?" Minn asked, her rifle at low ready.

"The primary chambers are forty meters below the surface," Boroleas said. "The excavation would have taken significant resources and time."

"Six weeks," Tessek said grimly. "About exactly the time I was mapping for them before I realized something was wrong."

They found the excavation shaft behind the largest intact wall of the ruin — a wide circular bore cut cleanly through the ancient stone and peat, fitted with a modern descent ladder of carbon-alloy rungs and lit by chemical glow-strips. It looked completely incongruous against the thousand-year-old stone.

"Old ruin, new hole," Prism said. "Metaphor for the whole of MarBryn, really."

They descended.

The passages below were a revelation — wide corridors of polished boron-ceramic tile, immaculate despite the millennia, their surfaces softly luminescent with the stored energy of the lattice. Alchemical script ran in continuous bands along the walls — not decorative, but functional, encoding the molecular stabilization equations of the Concordat's master engineers. Boroleas ran her gloved fingers along the script as they walked, reading it the way others might read a letter from an old friend.

"This is extraordinary," Minn said quietly, her tactical tension briefly overridden by genuine awe.

"The Concordat engineers were extraordinary," Boroleas agreed. "They understood boron chemistry at a level that current materials science is only beginning to approach. These tiles are still structurally perfect. The thermal conductivity has not degraded in a thousand years."

"Beautiful and also very dangerous," Tessek said, studying the junctions of the corridor walls. "This architecture is load-bearing on a molecular level. If someone runs a high-energy extraction sequence through this material—"

"It unravels," Boroleas said. "Like pulling a single thread from a precise lattice weave. Everything connected to it fails simultaneously."

They heard them before they saw them — the mechanical hum of extraction equipment running at capacity, and beneath it, voices. Boroleas pressed her group against the wall with a gesture, extinguished Prism's ambient glow with a look, and moved forward silently to the edge of a vast chamber.

The chamber was breathtaking and horrifying in equal measure. It was domed, perhaps thirty meters across, its ceiling a continuous arc of alchemical ceramic inlaid with boron-crystal matrices that glowed gold and blue. And it was full of Vorrath Keln's people.

She counted quickly: twelve operatives in composite armor, three engineers managing large extraction rigs, and at the center of it all, standing at the sealed door of the primary Vault with his hands clasped behind his back, Vorrath Keln.

He was taller than she remembered. Or perhaps that was the chamber. He was perhaps sixty, angular, with close-cropped white hair and the particular quality of stillness that belongs to people who are never, under any circumstances, surprised. He wore a long dark coat with the alchemical rank insignia of the old Morden Institute — a school Boroleas knew well, because she had been responsible for having it shut down.

"He looks," Prism breathed near her ear, barely audible, "like a man who has been planning this for a very long time."

"Nineteen years," Boroleas said quietly. "Since the last time I stopped him."

"What did he want then?"

Her violet-blue eyes were fixed on Keln. "The same thing he wants now. To unmake the world as it is and rebuild it according to his specifications." She paused. "He believes that If you control elemental composition at a deep enough level, you control everything."

"Does he have the knowledge to use the Primary Bindings?"

"He has most of it," she said. "And my notes. Which he stole twenty years ago."

Minn appeared at her shoulder. "So we have an extraction operation nearing completion, twelve armed personnel, three engineers, and one very dangerous alchemist, in a structurally sensitive chamber forty meters underground."

"Yes," Boroleas said.

"Good," said Minn. "I was worried this was going to be complicated."

Chapter 5: The Diamond Barrier and the First Strike

Boroleas did not rush. She never did. She stepped back from the chamber entrance and drew the rest of her group into a side passage, and for several minutes she simply thought — her gloved hands folded together, her eyes distant, moving through the problem the way she always moved through problems: systematically, from foundation to surface.

"If we engage directly in that chamber," she said, "the energy discharge risk to the boron-ceramic infrastructure is significant. I cannot fire energy bolts at full intensity in an enclosed space filled with structurally critical material."

"Constrained engagement," Tessek said, understanding immediately. He began sorting through the compact devices on his belt. "I have three EMP-adjacent pulse devices — they won't damage ceramic but will disable the extraction rigs and most of the communication technology the operatives are carrying."

"If I can get above the chamber," Minn said, studying a structural schematic Tessek had pulled from a folded survey map, "there are ventilation passages in the dome ceiling. I can use precision chemical rounds — non-explosive, gas-dispersal tipped. Incapacitating but structurally safe."

"Prism," Boroleas said.

"Misdirection and reconnaissance," the Luminal Fox said promptly. "I can refract my luminescence into targeted light patterns that will create significant visual confusion among personnel not trained to account for crystalline optics. Also, I can find and report alternate exit routes very quickly."

"What about Keln himself?" Tessek asked.

"Keln is mine," Boroleas said simply.

They moved.

Tessek took the left approach, Minn peeled away upward toward the ventilation shafts with practiced silence, and Prism slipped into the chamber like a moving prism of scattered light. Boroleas walked through the main entrance at a deliberate, unhurried pace, her magenta cape trailing, the blue-white orb already forming between her raised left hand.

She had covered perhaps five meters into the chamber before one of the armored operatives spotted her. The shout that went up was sharp and professional — these were trained people, not panicking. Three of them moved immediately to intercept, weapons raised.

Boroleas expanded the orb in a single smooth motion, pushing it outward and downward simultaneously. A dome of luminous boron-energy snapped into existence around her — a full protective barrier, curved and solid as forged steel, its surface humming with the frequency of crystallized boron molecules locked in a force lattice. Discharge bolts from two weapons struck it and diffused harmlessly.

At the same moment, Tessek's pulse devices detonated in sequence — three sharp, flat concussive sounds that sent the extraction rigs into immediate shutdown and caused half the operatives to stagger as their communication and navigation technology failed simultaneously. Minn's first dispersal round dropped through the ventilation gap above and burst in a cone of silvery incapacitating mist over four clustered operatives, who sat down heavily and stayed seated. Prism erupted from a side shadow in a blaze of fractured rainbow light, and three more operatives spun toward the visual chaos and found nothing to target.

It was, Boroleas reflected briefly, good teamwork.

Keln had not moved. He stood at the Vault door watching her approach with his hands still clasped, his expression that of a man watching something he had anticipated.

"Boroleas," he said. "I wondered how long."

"Six weeks, apparently," she said. "I should have been faster."

"The grid extraction is already ninety-three percent complete," Keln said. "Even if you stop the rigs, the energy is already routed into the Vault access mechanism. It will cycle fully in approximately forty minutes. You can't reverse it." His tone was not triumphant — it was simply informational. That was what made him so dangerous. He did not rage or boast. He calculated.

"Forty minutes," Boroleas said.

"Forty minutes," he confirmed.

Two remaining operatives came at her from the flanks. She sent a boron bolt sideways at low intensity — enough to impact the weapon from the first operative's hands — then pivoted and encased the second operative's boots in a rapid boron-hardened shell that locked him to the floor temporarily. Neither injured. Both neutralized.

"You were always precise," Keln said. "It was always your flaw and your virtue."

"Tell me about the Primary Binding sequence you intend to use," she said. "Walk me through it. Because I know the Vault's internal architecture better than you do, Keln, and there is something in your plan that you've miscalculated."

For the first time, the absolute stillness in his expression shifted — barely, almost nothing, but Boroleas was very good at reading faces.

She had him. She just needed time to figure out what the miscalculation was.

Chapter 6: Allies in the Deep

The battle in the chamber was effectively over, though it hadn't felt like a battle — more like a precise and organized dismantling. Minn descended from the ventilation passage via a carbon-line drop and began securing the incapacitated operatives with professional efficiency. Tessek moved through the chamber checking the extraction rigs with the expression of a physician examining a patient he had not arrived in time to save.

"The routing is completed," he said, his voice carefully neutral. "The energy is inside the access mechanism. He's right about that."

Prism was examining the Vault door beside Keln, who stood with two of Boroleas's boron-hardened restraint bands securing his wrists — not painful, not damaging, but absolute. He appeared to be thinking, which Boroleas found more concerning than anger would have been.

"What did you mean," Minn asked, pulling a restraint strap tight, "when you said he miscalculated?"

"I needed him talking," Boroleas said quietly. "And I needed time to think." She moved to the Vault door and pressed her gloved hand against its surface. The material sang to her — centuries of stored energy, boron-ceramic layers interlocked in configurations of extraordinary complexity. She closed her eyes.

"Can you feel what's been done to it?" Prism asked.

"Yes. The extraction energy has been loaded into the access sequence as a forced key — essentially using brute alchemical force where precision should be used." She paused, her eyes still closed. "Keln, you loaded the lattice energy into the primary access loop. But the Primary Binding sequence requires a secondary harmonic confirmation to activate. Without it, the forced key won't open the Vault. It will detonate it."

Silence.

"Specifically," she continued, "it will release forty meters' worth of stored boron-crystalline energy in an uncontrolled omnidirectional burst. The passage architecture would survive. The Vault contents would be destroyed. And everyone in this chamber would not." She opened her eyes and looked at him.

Keln's jaw was tight. "You're bluffing."

"I built the secondary harmonic lock," Boroleas said. "Twenty-two years ago, when I first sealed the Vaults. I installed it specifically because I was concerned about forced-entry attempts." She paused. "You're welcome."

The chamber was very quiet. Even the incapacitated operatives seemed to be listening.

A sound came from deeper in the passage — movement, multiple footsteps, not hostile in cadence. A moment later, two figures entered the chamber: a tall woman with bright copper hair pulled back severely, wearing a long coat bearing the insignia of the Alchemical Cooperative — MarBryn's interdisciplinary scientific governance body — and beside her, a young man of perhaps twenty with the wide, intensely curious eyes of someone who had never in his life stopped asking questions.

"Boroleas," the copper-haired woman said. "We've been tracking the lattice drain for two days. I am Alchemist-Surveyor Dara Venn, Cooperative Third Council. This is my research apprentice, Faldo. He is annoying but useful."

Faldo did not appear bothered by this description. He was already examining the extraction rigs with something approaching joy. "These use a boron-frequency resonance oscillator," he said to no one in particular. "That's theoretically brilliant. Also catastrophically irresponsible. Fascinating."

"The Cooperative sent you?" Boroleas asked Dara.

"I sent myself," Dara said bluntly. "The Council is still arguing about jurisdiction. I got tired of waiting." She surveyed the bound operatives and the frozen extraction rigs with a measuring look. "How much time before the access sequence triggers?"

"Approximately twenty-eight minutes now," Boroleas said.

"Then we reverse it," Dara said. "Show me the access mechanism internals."

"Can you reverse a forced-load boron-lattice sequence?" Tessek asked.

Dara looked at him. "Pathwright?"

"Tessek. Third rank."

"Have you ever seen a forced-load sequence before?"

"No."

"Neither have I," Dara said crisply, "but Faldo has been modeling them theoretically for three years and Boroleas built the lock, so between us we should manage." She turned to the door. "Faldo, equations."

Faldo immediately produced a handheld computational tablet from his coat and began pulling up molecular modeling schematics. "If we can introduce a reverse-frequency pulse into the loaded lattice energy — something tuned to the exact harmonic inverse — it should discharge the energy safely into the chamber walls rather than the Vault mechanism." He looked at Boroleas. "You'd have to hold a containment barrier around the entire door during discharge. To catch any overpressure."

"How large a barrier?" Boroleas asked.

Faldo calculated quickly. "Hemisphere, six meters radius, for approximately four minutes."

"Manageable," said Boroleas.

"The pulse emitter," Dara said, looking at Tessek. "Can any of your devices produce a harmonic-frequency pulse?"

Tessek was already examining his belt tools. "I have a resonance calibrator. It's designed for structural assessment, not energy manipulation."

"Boroleas can tune it," Prism said. "She can alter the conductivity profile of any boron-composite device by direct manipulation. If the calibrator has a boron-alloy transducer, she can reconfigure the output frequency."

Tessek handed over the device without hesitation. Boroleas took it, felt its internal composition through her gloves — the familiar recognition of boron-alloy components — and began, carefully and methodically, adjusting it at the molecular level. The device vibrated in her hands. Its output frequency shifted in a slow, rising arc.

Keln watched from across the chamber. He said nothing. His eyes followed Boroleas's hands.

"You know," Minn said to him, not unkindly, "most people who come up against her realize about here that they've lost."

"I don't lose," Keln said. "I recalculate."

"That's almost the same thing," Minn said. "But not quite."

Sixteen minutes.

Chapter 7: Fire Below the Earth

The preparation took eleven of the remaining sixteen minutes — time that felt both very long and very short in the glowing quiet of the Vault chamber.

Boroleas held the recalibrated resonance device and walked Faldo through the final frequency verification, the young apprentice cross-checking her molecular adjustments against his computational models with the focused intensity of someone who understood exactly what would happen if they got this wrong. Dara coordinated the positioning — operatives secured, everyone without a direct role moved to the farthest passage junction, Tessek stationed at the extraction rig controls ready to physically disconnect all remaining power conduits on Boroleas's signal.

Prism took up a position near Keln, who had been moved to the passage wall. The Luminal Fox said nothing, but his silver eyes did not leave the man.

"You seem very calm," Keln said to Prism eventually.

"I am approximately four hundred years old," Prism said. "Almost nothing seems worth panicking about at this point."

"You've been with her a long time?" Keln asked.

"Long enough," Prism said, "to know that whatever she is attempting right now will work."

Three minutes.

Boroleas stood at the Vault door, the recalibrated resonance device in her right hand, her left hand raised with the blue-white orb already expanding outward. She breathed steadily, feeling the familiar molecular hum of boron energy all around her — in the walls, in the door, in the loaded access mechanism thrumming with the stolen lattice power from Crystalmere's grid. She could feel each layer of it, the way a musician hears separate instruments within a full orchestra.

"On your call, Boroleas," Dara said from the passage mouth.

"Tessek, ready?" Boroleas asked without turning.

"Ready," Tessek said, his hands on the manual disconnect levers of the two remaining active extraction rigs.

"Faldo?"

"Frequency locked," Faldo said, his voice tight with concentration. "The inverse harmonic is set. You have a tolerance window of approximately eight seconds between activation and full barrier deployment. After that, the energy pressure will exceed what the device can maintain and you'll need the barrier fully established."

Eight seconds. Boroleas nodded.

She activated the resonance device.

The sound it produced was not loud — it was almost gentle, a low crystalline tone that resonated at a frequency just below comfortable hearing, more felt than heard. But its effect was immediate. The Vault door began to vibrate, its surface luminescence shifting from steady gold to a flickering, uncertain amber as the forced-load energy inside the access mechanism encountered the inverse harmonic and began to destabilize.

"Barrier," Faldo said quietly. "Now."

Boroleas pushed.

The protective dome that erupted from her left hand was the largest she had ever constructed — a full six-meter hemisphere of compressed boron-force lattice, its inner surface rippling with contained energy. She felt the weight of it immediately, like carrying something vast and trembling, the stolen lattice energy from Crystalmere pressing outward against her barrier in waves as it sought release.

The door screamed — a high metallic sound of molecular stress — and then the energy discharged.

It was, even contained, spectacular. Blue-white light filled the inside of the barrier like a captured star, blazing so intensely that the shadows in the chamber vanished entirely for several seconds. The pressure against the barrier was enormous and constant, and Boroleas locked her feet against the floor and held, her violet-blue eyes open, watching the discharge, monitoring the barrier's integrity at the molecular level, adjusting, compensating, holding.

Three seconds. Four. Five.

Minn, watching from the passage with Dara, said nothing. Her knuckles were white on her rifle.

Six seconds. Seven.

Eight.

The light collapsed inward, the pressure dropped, and the chamber exhaled. The Vault door stood intact, its surface now dark and completely inert — the forced-load energy fully discharged into the barrier and safely dissipated. The extraction rigs were silent. Tessek had disconnected them at exactly the right moment.

Boroleas lowered her hand. The barrier dissolved. She stood very still for a moment.

"That," said Faldo into the silence, "was the single most impressive application of boron-force manipulation I have ever witnessed."

"It was the only application of boron-force manipulation you have ever witnessed," Dara said. But her voice was not entirely steady.

Prism padded across the chamber to Boroleas's side. "Well done," he said quietly.

"The door?" she asked.

"Intact," Tessek confirmed, moving quickly to check. "Vault contents are undisturbed. The forced-load mechanism is completely neutralized. We'd need to rebuild the access sequence from the original Concordat schematics to open it now, and that would take — months, at minimum."

"Good," Boroleas said. She turned and looked at Keln.

He was watching her with an expression that was not defeat, exactly, but was at least the recognition that his calculation had been wrong. He said nothing.

"Take him up," Boroleas said to Minn. "His engineers and operatives as well. The Cooperative can coordinate with Intelligence for custody and processing."

Minn began organizing the operatives' removal with characteristic efficiency. Dara was already on a communication device, speaking rapidly in the clipped, authoritative tones of someone who had been waiting for weeks to make exactly this call. Faldo was taking detailed scans of the Vault door's molecular structure with barely contained scientific delight.

Tessek appeared at Boroleas's shoulder. "It's not over, is it," he said. It was not quite a question.

"Keln has been building toward this for nineteen years," she said. "He will not stop because one approach failed." She looked at the dark, inert door. "But the Vaults are safe. And for now, that is enough."

She reached into her robe and touched the blue gemstone brooch at her chest briefly — a private gesture, a habit from long ago. Then she turned and walked toward the passage and the long climb back to the surface and the light.

Chapter 8: The Surface World Answers

Emerging from the excavation shaft into Ashenmoor's ruined courtyard felt like surfacing from deep water. The overcast sky was brilliant compared to the glow-strip light below, and the air — metallic and damp — tasted extraordinary simply by virtue of being open and moving. MarBryn's moon was visible through a gap in the cloud, pale and small in the afternoon sky.

Minn oversaw the transfer of Keln and his twenty-three personnel to a Cooperative rapid-response unit that arrived by steam-turbine vehicle convoy forty minutes after Dara's call — three armored transports and a mobile command unit that unfolded from a flatbed trailer into a functioning field station with the smooth mechanical choreography of a well-designed machine. The Cooperative officers were professional and took custody without complication. Keln walked to the transport without resistance, without a word, and without once looking back.

Boroleas watched the convoy depart. Prism sat beside her.

"He'll be out within two years," she said. "The Cooperative's legal framework isn't built for someone like him."

"Then we stay alert," Prism said simply.

Dara Venn emerged from the mobile command unit with two cups of hot chemical-brewed tea — a pragmatic woman's response to the aftermath of an underground crisis — and offered one to Boroleas, who accepted it with genuine gratitude.

"The Council will want a full report," Dara said. "Specifically, they'll want to know why there is a functioning alchemical Vault of extraordinary strategic significance beneath a public ruin that was apparently not on any monitored list."

"Because I removed it from the monitored list," Boroleas said.

Dara regarded her. "You removed a primary strategic asset from the Cooperative's security registry."

"I removed it from every registry. The safest lock is the one no one knows exists." She drank her tea. "That philosophy served well for nineteen years. It only failed because I designed the secondary harmonic with a key that left a mathematical signature Keln eventually identified in the stolen notes."

"The notes you mentioned earlier," Dara said. "He stole them from you."

"Twenty years ago. A long operation, very carefully executed. It took me three years to realize what precisely he had taken." She paused. "Those notes were the only record of the harmonic lock's design. I should have burned them. I was sentimental about my own work."

"Aren't we all," Dara said.

They sat with that for a moment.

Faldo appeared from the direction of the excavation shaft, slightly dirty, apparently having gone back down for additional scans immediately after the convoy departed. He was vibrating with the contained energy of someone with six important things to say and insufficient social calibration to determine which to say first.

"The boron-ceramic tile work in the lower passages," he began immediately, "uses a molecular interlocking pattern that is twenty percent more efficient than our current best fabrication methods. The Concordat engineers were using a sintering technique we haven't independently replicated. I took full spectral scans. I think if we analyzed the crystalline boundary structures—"

"Faldo," Dara said.

"Yes?"

"Eat something first."

He looked down, apparently noticing for the first time that he was holding his scanning tablet with both hands and had no food. "Right," he said. "Yes. After."

Tessek had been working quietly at the perimeter of the ruin since they had emerged, moving methodically through the outer structures with his Pathwright's trained eye, touching walls, examining foundations, occasionally making notations on a small engineering pad. He returned to the group now and crouched to examine something in the courtyard ground.

"There is a secondary excavation bore," he said. "Smaller. Partially concealed." He pointed to a section of collapsed wall. "It was dug and then deliberately disguised after the fact. Backfilled with rubbed stone aggregate to match the surrounding material." He looked up. "Keln's team didn't dig this one. The tool mark signatures are completely different. Different equipment, different time frame — this one is significantly older."

"How old?" Boroleas asked, moving to look.

Tessek measured the oxidation on the bore edge carefully. "The chemical weathering profile suggests — ten, twelve years ago at the most recent, but the fill material itself is older than the main excavation." He paused. "Someone else was investigating the Vaults before Keln."

The group looked at each other.

"Who else knew they existed?" Minn asked, looking at Boroleas.

"The six founding houses," Boroleas said slowly. "All of whose documented lines I believed extinct." She turned to face the disguised bore entrance, her eyes very bright. "Apparently I was wrong about that as well."

Faldo raised one hand tentatively. "Does this mean we are immediately going into another underground passage?"

"Not tonight," Boroleas said. "Tonight we rest and resupply." She looked at the group assembled around her — Prism, Tessek, Minn, Dara, and Faldo — and felt the particular kind of warmth that comes from trust earned in difficult circumstances. "All of you. You fought well today, and you thought clearly. I am grateful."

"I didn't actually fight," Faldo said.

"You saved the Vault," Boroleas said. "That counts."

His expression was something between surprised and deeply pleased. Dara gave him a look that suggested she agreed but would never say so directly, which, Boroleas suspected, was Dara's way of saying it.

Chapter 9: The Sixth House

They camped that night on the high ground above the Ashenmoor ruins, in the shelter of a basalt outcropping that Tessek identified as structurally sound in under four minutes. A small thermal element provided warmth — modern, compact, boron-nitride insulated — and the group gathered around it as MarBryn's moon rose, casting the wetlands below in overlapping pale and rust-colored light.

Dara and Faldo needed to return to the Cooperative headquarters the following morning — there were reports to file, evidence chains to establish, and the considerable administrative task of explaining to the Council why its Third Alchemist-Surveyor had gone into the field unilaterally and returned with twenty-three prisoners and a revelation about an unregistered strategic Vault. Dara was composed about this prospect in the way that people are composed about things they have already decided to face without apology.

"I'll return when I can," she told Boroleas before sleeping. "Faldo will as well, if he's any use to you."

"Constantly useful," Boroleas said, which was true.

"He knows," Dara said quietly. "That's part of why he never stops."

In the early morning they said their farewells — Faldo nearly forgot his tablet in his eagerness to promise a full molecular analysis report within the week, recovered it, shook everyone's hands, and then departed beside Dara in the Cooperative's remaining transport with the distinctive energy of a person heading toward a great deal of very interesting work.

Boroleas, Prism, Minn, and Tessek descended into the ruins at first light.

The concealed bore was exactly as Tessek had marked it — a tight passage, barely wide enough for a person to move through sideways at points, running not down toward the Vaults but laterally, threading through the ruins' foundations at a shallow angle heading northeast. The walls of the bore were smooth, and here and there they encountered something that made Boroleas slow and look carefully: tiny alchemical sigils, scratched into the stone at intervals, in the specific script of the founding Concordat houses.

"Breadcrumbs," Minn said, holding a glow-strip to the wall.

"Guidance marks," Boroleas said. "Someone was here who understood the Concordat system. Who wanted to be able to find their way back." She traced the sigil with one gloved finger without touching it. "This symbol — this is the mark of the Sixth House. Ceraleth."

"I thought the six houses were extinct," Tessek said.

"The House itself is. The lineage of practitioners..." She paused. "Less certain."

The passage ended after approximately two hundred meters in a carved alcove, where they found a sealed message container of boron-ceramic construction — small, palm-sized, locked with a molecular key that Boroleas could feel the moment she picked it up. It was tuned specifically to boron-energy manipulation. It was, unmistakably, made for her.

She unlocked it with the lightest touch of her awareness.

Inside was a single thin plate of alchemical crystal, inscribed with dense text. Boroleas read it standing in the alcove while the others waited. Her expression shifted through several things — recognition, grief, surprise, and finally a kind of resolute clarity.

"What does it say?" Prism asked.

"It's from a woman named Sela Ceraleth," Boroleas said. "She was the last documented member of the Sixth House lineage. She died eleven years ago." She looked up. "She knew about Keln. She had been tracking his investigation into the Vaults for over a decade before any of us were aware of it. She left this here because she believed that whoever eventually came to deal with Keln would need to know what she had discovered."

"What did she discover?" Minn asked.

"That the Primary Bindings in the Vault are only half the mechanism," Boroleas said. "The other half — the actuating component that makes the Bindings functional — is not in the Vault." She held up the crystal plate. "Sela found it and moved it. Eleven years ago, she removed the actuating component from the Vault and hid it somewhere else. That's why she came here — not to open the Vault, but to ensure that even if someone did, they could never use what they found."

"She protected it," Tessek said.

"She did. And she left coordinates." Boroleas looked at the final lines of inscribed text. "The actuating component is hidden in a place called the Vitreous Shelf — a location in the high country above the Aetherveil Mountains."

"That's four days' travel north," Tessek said immediately.

"Then we go north," Boroleas said. She looked at the crystal plate for a moment longer. "Sela Ceraleth worked alone for eleven years to protect something that most people didn't know existed. She never told anyone." She was quiet for a moment. "I wish I had known her."

"She clearly knew you," Prism said softly. "The container was made for your hands."

Boroleas closed her fingers around the crystal plate carefully. "Yes," she said. "It was."

They returned to the surface and turned north, into the wind coming down from the Aetherveil peaks, cold and clean and carrying the high, sharp scent of elevation. The basalt plains gave way gradually to rising ground — scrub and stone and the occasional ancient waymarker, carved in the old Concordat style, that most travelers passed without recognizing for what they were.

On the second day north, Minn took a long-range shot from a concealed position on a ridge above them — the discharge weapon crack barely audible across the distance. She dropped flat as a second shot came in, rolled behind a stone shelf, and came up with her rifle tracking the ridge line.

"Sniper," she said calmly. "Single shooter. Ridge at two hundred meters, northwest."

Boroleas had already extended a partial boron barrier over the group — a low angled deflection shield, sufficient for the current threat. "Keln's people?" Tessek asked, pressing against the stone.

"Keln is in Cooperative custody," Minn said, sighting carefully. "These are someone else's people."

She fired twice — chemical propellant rounds, precise — and there was a distant sound of someone losing their footing on loose stone and a weapon clattering. "One down. Relocating." She was already moving.

Boroleas looked at Prism. "How many?"

Prism had risen to his full crystalline luminescence, his light scattering in deliberately confusing patterns. "Three total on the ridge. Two more in the shallow gully to the east. Five."

"Five and a sniper position suggests a professional intercept team," Tessek said, his own compact device — now reloaded — in his hand.

"Sent by Keln before his capture," Boroleas said. "A contingency. Ordered before he was taken." She stood, let the barrier expand from defensive to active, and sent two precisely calibrated boron bolts toward the gully to the east in rapid succession. The bolts struck the ground before the concealed figures, detonating in flashes of vivid blue that scattered stone and sent both individuals scrambling away in retreat. "They want the crystal plate," she said. "They know about Sela's message."

The skirmish lasted eight minutes. It was fast, violent in a contained way, and ended without deaths — three interceptors fled and two were neutralized and restrained by Minn, who had an extremely practical approach to all forms of tactical problem. Tessek's arm was grazed by a ricochet that left a significant burn, and he sat with characteristic stoicism while Minn applied field-grade chemical treatment to it.

"That's the second time I've been injured on this expedition," he said.

"You're maintaining an impressive average," Prism said.

"Thank you," Tessek said, though he did not sound entirely as though he meant it.

They pressed on north. The Aetherveil peaks rose ahead of them, white and ancient, their slopes gleaming in the late afternoon light. Somewhere in the high country above the frost line, in a place called the Vitreous Shelf, the last piece of an ancient and terrible mechanism waited in the dark.

Chapter 10: The Vitreous Shelf

The Vitreous Shelf was not on any modern map. Tessek — who had memorized more geological surveys than he generally admitted to — identified it from the Concordat waymarkers that appeared with increasing frequency as they climbed through the lower Aetherveil passes, each carved stone a breadcrumb in Sela Ceraleth's eleven-year trail. On the fourth morning the trail took them above the frost line, through a narrow defile of pale grey stone that opened suddenly onto something that stopped all of them in their tracks.

It was a plateau of extraordinary beauty. The stone of the shelf was fused silicate — vitreous, as its name described — a broad expanse of naturally glass-smooth rock formed in some geological ancient event that had melted and cooled the surface into a vast mirror of pale transparency. Beneath it, trapped in the frozen glass of the rock, were visible layers of mineral deposit in banded colors: amber, deep green, white, faint violet. MarBryn's morning light caught the surface and scattered it into a thousand quiet prismatic fragments across the snowfields at the plateau's edge.

Prism sat down and looked at it for a long moment in silence. Coming from him, silence was a form of reverence.

"My cave system," he said eventually. "In the frost peaks. It looked something like this. When I was very young."

"How young is very young?" Minn asked.

"Perhaps thirty or forty," Prism said. "I was reckless and fast then. I hadn't learned to observe properly yet."

Tessek smiled despite his bandaged arm. "I think I'd have liked you at thirty."

"I was considerably less articulate," Prism said.

Boroleas was already moving across the shelf, reading the ground with her awareness extended — feeling through the vitreous stone for the particular molecular signature of boron-ceramic work, the unmistakable craftsmanship of Concordat construction. She found it near the center of the plateau: a sealed depression in the rock, barely visible on the surface, its edges traced with Sela's Sixth House sigil marks. She pressed her gloved hand to the stone and felt the lock — a boron-energy key, identical in design to the message container.

"This was made by the same hands," she said quietly.

She opened it.

The depression unsealed with a sound like a single clear musical note — a harmonic release that sent a brief visible ripple through the glassy stone surface in all directions, as though a stone had been dropped in still water. Within the revealed hollow, resting in a cradle of boron-ceramic insulation, was the actuating component.

It was smaller than Boroleas expected — a crystalline lattice structure roughly the size of her fist, its interior a complex three-dimensional arrangement of boron and carbon matrices that glowed faintly gold in the morning light. It was, in its way, extraordinarily beautiful. It was also the most dangerous object in MarBryn.

"That's it," Minn said.

"That's it," Boroleas confirmed.

She lifted it carefully, feeling its molecular structure with her awareness as she did — reading it the way she read all things made of boron, thoroughly and completely. What she found inside it made her go still for a long moment.

"What is it?" Tessek asked, watching her expression.

"It isn't only the actuating component for the Primary Bindings," she said slowly. "That's what Sela believed it to be. That's what the Concordat records said it was." She turned the lattice structure carefully in her gloved hands. "But there's a second function encoded in the molecular architecture. A constructive function, not a destructive one." She looked up. "The same mechanism that could be used to unmake physical structures — the same Primary Binding system — could be used to build them. To repair them. To restore molecular integrity to damaged or degraded alchemical infrastructure." She paused. "To restore a city's boron lattice grid, for example."

No one spoke.

"The Concordat built it as a repair tool," she said. "A means of maintaining the great alchemical works they had built across MarBryn. The destructive application was always secondary — a side effect of a technology designed for construction and care." She was quiet for a moment. "Keln saw only the weapon. Sela saw only the danger. Neither of them read it all the way through."

"You read it all the way through," Prism said. "In thirty seconds, holding it for the first time."

"It's built from boron," Boroleas said simply. "I know the language."

The morning light on the Vitreous Shelf was warm now, and the scattered prisms from the glassy stone danced across all of them — across Minn's composed, watchful face, across Tessek's calloused hands resting at his sides, across Prism's crystalline form that caught and remade the light as it always did.

Boroleas held the lattice structure and made her decision.

"The Crystalmere grid was drained by Keln's extraction operation," she said. "Ninety-three percent of the stored energy is gone. Without restoration, the grid will degrade within five years, and when it fails, the boron-ceramic structural integrity of every building in the city goes with it." She looked at each of them in turn. "I can use this to restore it."

"Will that neutralize it?" Tessek asked. "The destructive potential?"

"No. Not permanently. But if I use the constructive function first — pour the actuating energy into a restoration sequence rather than a destructive one — the residual molecular state will be constructive. It would take years of reconfiguration to reverse it back toward the destructive application." She considered. "Years, and my notes, which Keln no longer has access to."

"And after you restore the grid?" Minn asked.

"I dismantle the actuating component," Boroleas said. "Carefully. Dispersing its molecular matrices into the grid itself, distributing the energy through a system large enough that it becomes part of the infrastructure rather than a separate, portable weapon." She paused. "The Concordat built it to be used and integrated. That's what it was always designed for."

"Is that — can you do that?" Tessek asked. "Just decide to dismantle something the Concordat spent years building?"

"The Concordat no longer exists," Boroleas said. "The houses are gone. Sela Ceraleth is gone. What remains is the responsibility to use what they left wisely." Her violet-blue eyes were clear and steady. "I've been a guardian of this knowledge for sixty years. This is what guardianship means."

Minn unfolded her arms. "Then let's go back to Crystalmere."

The return journey took two and a half days and passed without significant incident — though on the last afternoon, as they descended through the Thornfeld Plains, Faldo reappeared on the road ahead of them, slightly out of breath, carrying his scanning tablet and a leather satchel that clinked with vials and instruments.

"Dara said I could come back," he said, falling into step beside them. "She's filing the final report herself. She's better at administrative language than I am — I tend to write equations in the margins."

"She also sent supplies," Prism observed, eyeing the satchel.

"Mostly," Faldo said. "Some of it is my instruments. The molecular scan data from the Vault passages is going to take months to fully analyze and I was hoping—" He stopped, apparently remembering social context. "Is everyone well? Tessek, is that a significant injury?"

"It's managed," Tessek said.

"Good. Excellent. The scan data really is—"

"Walk and talk," Minn said.

He did.

They reached Crystalmere the following morning. The city was busy and unchanged on its surface — trams, vendors, gas lamps, the daily industry of a place that did not yet know how close it had come to catastrophe. Boroleas walked through the market district where she had first felt the grid's disturbance, and she found a quiet corner of the central square — a space between an old fountain and a newer communication relay tower, where the cobblestones above the lattice were oldest and thickest.

She pressed her palm flat to the stone and went deep.

The restoration took four hours. To anyone watching, it would have appeared that an alchemist in a magenta cape was simply sitting in the square, perfectly still, her gloved hand on the cobblestones and a faint glow emanating from between her fingers. Minn kept the curious at a respectful distance with a combination of polite authority and the sort of look that communicates, without speaking, that questions would not be welcome at this time. Tessek monitored the lattice resonance with his recalibrated devices and reported steady improvements in measured intervals. Faldo took scans of everything with breathless enthusiasm and whispered commentary that Prism patiently absorbed.

The actuating component dissolved into the grid slowly — its molecular lattice unraveling into the boron-ceramic network with the careful, unhurried precision that Boroleas applied to all things. She did not rush it. She felt each thread of it integrate into the city's architecture, becoming part of the structural foundation rather than separate from it. The energy distributed through the system like warmth spreading through cold stone — gradual, thorough, and permanent.

When it was complete, she lifted her hand.

The square looked exactly the same. The trams ran. The vendors called their prices. An old man near the fountain dozed in the morning sun. But beneath all of it, the boron lattice sang at full strength for the first time in weeks — restored, reinforced, and woven through with something more than it had been before. Sixty years of

Boroleas's accumulated alchemical understanding, poured into the infrastructure of a city that would never know her hands were in its foundations.

She stood, brushed the stone dust from her glove, and looked up at the sky over Crystalmere — wide and blue and ordinary and exactly as it should be.

"How do you feel?" Prism asked.

"Tired," she said. "Satisfied." She considered. "Hungry."

Prism's silver eyes brightened with something that, on a more conventional creature, would have been called a smile. "Alderman Petch's neighbor two stalls down sells hot meat pies," he said. "I noticed on the way in."

"You don't eat," Boroleas said.

"No," he agreed. "But I notice things."

The group came together at the fountain — Minn, Tessek, Faldo, and Prism, in the ordinary amber light of mid-morning Crystalmere — and Boroleas looked at each of them. Minn, who had walked into an underground operation without being asked and handled every moment of it with the contained competence of someone twice her age. Tessek, who had been hired by the enemy, recognized the wrong in it, and then sat bleeding on a causeway and chosen the right side without making a production of it. Faldo, who would spend months extracting extraordinary knowledge from spectral scans and would write it all in the margins of reports that Dara would patiently reformat. And Prism, four hundred years old and silver-eyed, who had been at her side long enough that she had stopped trying to calculate precisely how long.

"The Cooperative will want debriefs from all of you," Boroleas said. "Keln will face formal proceedings. The Vault will need to be properly assessed and registered. There is a great deal of administrative work ahead."

"Thrilling," Minn said drily.

"There will also," Boroleas said, "be a very good meal first. On me." She paused. "Prism excluded, since he subsists on ambient light and refuses to be practical about it."

"I am extremely practical," Prism said. "I simply don't require food. It's efficiency, not exclusion."

Tessek laughed — a full, genuine sound that seemed to surprise him slightly. Faldo was already writing something in the margin of his tablet. Minn looked at the hot food vendor down the square with an expression that was, for her, practically enthusiastic.

Boroleas pulled her magenta hood up against the morning breeze, the blue gemstone brooch catching the light at her chest, the golden sigils on her robe warm in the sun. The golden staff at her back stood steady. Her violet-blue eyes moved across the city she had helped hold together — the old stones and new trams and ceramic buildings that would stand, now, for another generation at least.

Crystalmere would be all right.

For today, that was enough.

She walked forward, her cape billowing behind her in the way it always did — as though even the air understood that she was moving toward something, and chose to follow.

Boron Facts

- Pure Boron first produced in 1909 in USA
- Solid alkaline metal
- Good conductor of heat
- Poor conductor of electricity
- Diamagnetic
- Brittle metal only malleable when hot
- Metalloid

Author's Note

Every great story begins with a spark of curiosity — and this one began with a tiny, remarkable element sitting quietly in the fifth position on the periodic table.

Boron has always fascinated me. It is, in many ways, a paradox made physical: brittle yet extraordinarily strong in compound form, humble in appearance yet essential to some of the most advanced materials humanity has ever created. Boron carbide is used in bulletproof armor. Borosilicate glass withstands extreme thermal shock. Boron nitride rivals diamond in hardness. And yet, for all its remarkable qualities, Boron rarely gets the recognition it deserves.

When I first imagined Boroleas — her crackling energy bolts, her luminous crystalline barriers, her precise and calculating mind — I was asking a question that genuinely excited me: What would it look like if the true properties of boron became the foundation of a magical system? What if the lattice structures, the hardness, the heat resistance — what if all of that became power? What if chemistry itself was the source of magic?

The chemistry in these pages is real at its roots. The boron-ceramic technology, the lattice grid structures, the molecular manipulation — these concepts are grounded in genuine science, stretched into fantasy, but never invented from nothing. My hope is that a curious reader who encounters boron carbide in these pages will one day look it up — and feel that same electric thrill of recognition when they discover it is real.

That is the heart of this series: the belief that science does not diminish wonder. It creates it. The periodic table is not a dry list of abbreviations. It is a map of everything that has ever existed — and every element on it has a story worth telling.

To every reader who has ever felt that science was too distant to love: I wrote this for you.

The periodic table has one hundred and eighteen stories waiting to be told. This is just one of them.

Science has never felt more magical. Magic has never felt more real.

Thank you for joining me in this journey,

Sybrina Durant

PS: If you're interested in reading the resources the scientific data in this book was derived from, check out this webpage: ***https://magicalptelements.com/mw1pt/***

About The Author

Sybrina Durant is the author of

"The Magical Elementals Chemistry Fantasy Series"

"The Magical Elements of the Periodic Table Series"

"The No Metal No Magic Series"

"The Blue Unicorn – Journey To Osm" books

and she is the creator of the "Elemental Fantasy Periodic Table" which is available as a brilliantly colored poster.

There are also book activities, songs, trading cards and t-shirts for all 118 elements and elementals, too.

Visit MagicalPTElements.com to see everything that is currently available.

If you love to sing and learn, check out the "Chemistry Sing Alongs" playlist on the Sybrinad channel on Youtube.

Glossary of Scientific Terms in Boroleas the Alchemical Wizard with Boron Magic

Boron (Element 5) - A real chemical element prized for its lightness, exceptional strength, high heat resistance, and unique ability to form strong crystalline molecular networks (lattices). Boroleas's entire power set is drawn from this one element: her energy bolts, protective barriers, and lattice-sensing awareness are all forms of boron manipulation. It was her heightened, almost musical sensitivity to Element 5 that let her detect "a trail of destabilized boron molecules" drifting through the Crystalmere market before anyone else noticed anything was wrong, and it is the same element she helped weave into the city's foundation nearly sixty years earlier.

Boron Lattice / Lattice Grid - A highly organized, repeating three-dimensional geometric arrangement of boron atoms linked together; in real science, lattice structures give materials immense strength. In the story, this is the literal foundation of Boroleas's magic — the underground network of boron-ceramic conduits beneath Crystalmere that she helped the Concordat engineers build and stabilize decades ago. When Vorrath Keln's operatives illegally drain the grid, buildings crack with no impact point, and it is Boroleas who ultimately kneels in the city square for four hours, pouring her own molecular awareness back into the lattice to restore it thread by thread.

Boron Ceramic - A dense, heat-resistant compound made by combining boron with other elements such as nitrogen or carbon. In the story, boron ceramic is the Concordat's signature building material — used for the city's structural tiles, the corridors beneath Ashenmoor, and the sealed Vault door itself. Boroleas can "read" boron ceramic the way others read a letter, running her gloved fingers along it to sense stored energy, hidden fractures, or tampering, which is exactly how she diagnoses the sabotage at the heart of the plot.

Resonant Frequency Disruption - A phenomenon in which specific energy vibrations match the natural frequency of a solid structure, causing it to shake, weaken, or fracture. This is the very clue that launches Boroleas's investigation: ceramic display cases in the market crack overnight with "no impact," which Prism immediately identifies as the signature of an unauthorized extraction frequency being run through the grid — proof that someone is deliberately weaponizing boron science against the city Boroleas swore to protect.

Thermal Conductivity - The measure of how easily heat energy moves through a material; boron compounds naturally have high thermal conductivity, letting them withstand or transfer enormous thermal energy without melting. In the ruins beneath Ashenmoor, Boroleas marvels that the Concordat's boron-ceramic tiles have not degraded in a thousand years, a detail that grounds her magic in genuine materials science and foreshadows just how much raw energy the ancient lattice can safely hold — and how catastrophic it is when that energy is forced loose.

Molecular Manipulation - The ability to control, reorder, or alter the arrangement of individual molecules within a material. This is Boroleas's core magical discipline, used throughout the book in small, precise acts: locking Drevon Salk's hidden panel shut with a thought, hardening an operative's boots into the floor, fusing a weapon barrel solid, and — most critically — retuning Tessek's ordinary resonance calibrator at the molecular level so it can emit the exact inverse-harmonic frequency needed to save the Vault.

Boron Nitride / Boron Carbide - Advanced boron compounds: boron nitride rivals diamond in hardness and endures extreme temperatures, while boron carbide is extremely tough and light (used in real life for body armor). In the story, the boron nitride rods in Salk's workshop show telltale "micro-fracturing" from illegal energy extraction — the physical evidence that first exposes his operation — while boron-nitride-insulated equipment, like the group's camp thermal element, shows how the same real-world compound quietly powers everyday alchemical technology in Boroleas's world.

Harmonic Lock / Inverse Harmonic - A security or stabilization mechanism based on wave frequencies; an inverse harmonic emits an opposite energy wave to safely cancel out destructive built-up pressure. Twenty-two years before the story begins, Boroleas herself built the secondary harmonic lock protecting the Concordat Vault — the safeguard that stops Keln's forced entry from succeeding. Later, working with Faldo's calculations, she uses an inverse harmonic pulse to defuse the stolen lattice energy safely, turning a defensive concept from her own past into the tool that resolves the story's central crisis.

Destabilized Molecule - A molecule whose internal chemical bonds are strained or broken, making the surrounding substance weak, unpredictable, or unsafe. The opening chapter hinges on this concept: Boroleas is drawn into the entire plot because she senses "a trail of destabilized boron molecules" beneath the market square, a disturbance only someone with her particular sensitivity to boron magic could have detected at all.

Actuating Component - The physical or molecular "trigger" mechanism that initiates a complex chemical or alchemical process. In the story it is the fist-sized crystalline lattice object hidden for eleven years by Sela Ceraleth at the Vitreous Shelf — the missing half of the Primary Binding mechanism. Boroleas discovers it holds two opposite functions, destructive and constructive, and chooses the constructive path, using it to restore Crystalmere's boron lattice before carefully dismantling it into the grid for good — the story's clearest statement that Boroleas's magic is fundamentally about guardianship and repair, not destruction.

Primary Binding(s) - Described in the story as "the alchemical equivalent of a master code" — a sequence of elemental configurations that, properly understood and applied, allows its holder to alter the fundamental molecular composition of large areas of the physical world. The Primary Bindings are the ultimate expression of boron-based molecular manipulation in the series, and the entire plot turns on who controls them: Vorrath Keln wants to weaponize them to "unmake" the world, while Boroleas, once she recovers the actuating component, reveals their original, intended use was constructive — to build and repair, not destroy.

Sintering - A real materials-science process in which a compound is heated and compressed until its particles fuse into a single solid form without fully melting — commonly used to manufacture advanced ceramics. Faldo identifies the Concordat's boron-ceramic tiles as the product of "a sintering technique we haven't independently replicated," twenty percent more efficient than modern fabrication methods, underscoring just how far ahead of current science the Concordat engineers — and by extension Boroleas's own boron magic — truly were.

Vitreous / Fused Silicate - "Vitreous" describes a smooth, glass-like material formed when silicate rock is heated to extreme temperatures and rapidly cooled, fusing its mineral layers into a transparent or semi-transparent solid. The Vitreous Shelf, the remote plateau above the Aetherveil Mountains where Sela Ceraleth hid the actuating component, is named for exactly this geology — its glass-smooth stone is real-world silicate chemistry standing in for the story's most sacred hiding place, chosen because its molecular signature was subtle enough that only someone attuned to boron, like Boroleas, could find the sealed boron-ceramic depression within it.

Electromagnetic Pulse (EMP) - A burst of electromagnetic energy capable of disabling electronic and communication equipment without a physical explosion. Tessek's "EMP-adjacent pulse devices" are calibrated specifically not to damage boron ceramic while still disabling the extraction rigs and communications gear of Keln's operatives — a practical, non-magical technology that works alongside Boroleas's molecular magic during the chamber assault, illustrating how the story blends real engineering with alchemical power.

www.ingramcontent.com/pod-product-compliance
Lightning Source LLC
LaVergne TN
LVHW050747061126
851857LV00013B/1034
9781942740742